FILM & TV

BY CHRIS OXLADE

WHAT ARE FILM & TELEVISION?

Film and television are similar yet different media for showing moving pictures. Films are prerecorded, both on sets and in real locations. They usually tell an entertaining story and are shown in movie theaters. Television pictures are often broadcast into our homes "live," as the televised event (such as a football game) is actually happening. Such events are also recorded for broadcast at other times. Television has become one of our major sources of news, information, education, and entertainment. Filmmaking is just over 100 years old, and television about 30 years younger. Until the 1990s, only old films were shown on television, but now there are television channels that specialize in showing recently released films.

BIG BUDGETS

Some very successful films are low-budget movies made on simple sets with unknown actors — Hitchcock's *Psycho* (1960) was an early example. Others, such as *Star Wars: Phantom Menace* (left), cost millions of dollars to achieve complex action, special effects, and to pay big-name actors to star in them. The money spent to make these films is recovered from box-office receipts and sales of merchandise such as toys, T-shirts, posters, books, and videos.

MOVIEMAKING

Filmmaking is a huge industry that employs many thousands of people to make hundreds of new films every year. The most famous filmmaking center is Hollywood, a suburb of Los Angeles, where films have been made since about 1910. Here, on the set of the James Bond movie *Goldeneye*, a camera is mounted on the front of a car to film its driver.

TELEVISION STATIONS

There are hundreds of television stations around the world, many of which broadcast on several channels at once. They produce news, information, entertainment, and educational programs. Major events are broadcast live on television around the globe to millions of viewers. Some stations are operated by national broadcasting services, such as the BBC in Britain. Commercial stations make money by charging business and manufacturing companies a fee to broadcast their commercials or by charging the viewer to watch.

WHAT IF FILM & TELEVISION HAD NEVER BEEN INVENTED?

How many hours of television do you watch each day? Do you think you watch too much? How often do you go to the movies or watch videos? Imagine if film and television had never been invented. The main sources of information would be newspapers, the radio, books, and other people, just as it was until the 1950s when television became generally available. Like millions of other people around the world who live without television, we would find different ways to entertain ourselves.

GROWING UP WITH TELEVISION

Ninety percent of homes in the developed countries of the world have a television. Most children start watching television when they are very young and will watch it almost every day for the rest of their lives. When no suitable programs are being shown, there is often the option of watching a video instead.

ORIGINS OF MOVING PICTURES

U ntil 1840, very few people had seen a photograph, let alone a film. But they might have seen one of the optical devices (often toys) that were popular at the time. Such devices were the first stage in the development of motion pictures. A moving picture is made up of many pictures, called frames, each slightly different than the one before it. When these images are viewed in quick succession, our brain perceives the changing image as if it is moving. The effect relies on the fact that the eye remembers each picture for a split second after it has gone. This is called persistence of vision. When the pictures appear at more than 15 frames a second, the movement looks smooth. You can find out more about this by trying the activity on page 17.

THE FIRST MOVIE CAMERA

The French scientist Etienne-Jules Marey (1830–1904) was interested in animal movement, especially the flight of birds. In 1882, he developed a camera that could take 12 images a second on a rotating glass photographic plate. This was the first camera capable of recording moving images.

HOW PHOTOGRAPHY WORKS

In traditional photography, images are recorded by light-sensitive chemicals contained in film. A camera records a photograph by exposing the film to light coming from a scene. The film must be kept in the dark until it is developed by using chemicals to make the photographs visible. Photographic prints are made from the film.

Light-sensitive film is loaded into camera.

Film is exposed to light from the scene.

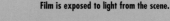

Film is developed using chemicals to make images show up on negatives.

Each photograph is printed from a negative.

THE FIRST MOVING IMAGE

This sequence of photographs was taken in the 1870s by Eadweard Muybridge (1830–1904) to record how animals move. The photographs were taken in quick succession by different cameras as the horse trotted in front of them. Muybridge used a device called a zoopraxiscope to display the photographs in a rapid sequence on a screen. This was the first time something real was shown as a moving image.

MAKING PICTURES MOVE

The first moving pictures were not films but cartoons drawn by hand. In the early 19th century, several different devices were invented to show these cartoons moving. In the praxinoscope, a set of slightly different pictures is arranged around the inside of a drum. As the drum spins, the viewer sees each picture for a split second in a mirror and is given the impression that the figure is animated.

EDISON'S KINETOSCOPE

The first camera to use roll film was developed in 1887, in the laboratories of the great inventor Thomas Alva Edison (1847–1931). It was called the kinetograph. A long roll of film was needed to record more than a second or so of motion. The films were shown on a device called a kinetoscope, which was like having a movie theater at home. A variation of the machine, the vitascope, projected the film onto a screen so that many people could watch at the same time.

THE FIRST MOVIES

Films, or "animated photographs" as they were called, were at first thought of as novelties like other optical toys. They were not films as we think of them today. The two main pioneers, Edison and the Lumière brothers, made very short films of circus and music hall acts, and of real-life events such as horse-drawn carriages moving along city streets. People were more fascinated by the pictures moving rather than by what was happening in them. Gradually, traveling entertainers bought film projectors and wanted a supply of new films to show to their audiences. The serious art of filmmaking began and soon permanent theaters were being built—the first by the Lumière brothers in 1895. The first of the famous Nickelodeon theaters opened in Pittsburgh, Pennsylvania, in 1905.

A CLASSIC FILM

In the first 20 years of the 20th century, the art of film-making became well established. Directors and cameramen developed techniques that added dramatic impact to their films. For example, *Battleship Potemkin* (1925) featured many short shots, each just a few seconds long, in its portrayal of a naval mutiny. It still ranks as one of the greatest films ever made.

THE TRAMP

British-born Charlie Chaplin (1889–1977) was one of the earliest stars of silent films. In his first film, *Making a Living* (1914), he played a tramp with a bowler hat, baggy trousers, and a cane—a character who was to become world famous. Among his most famous films are *The Gold Rush* (1925) and *Modern Times* (1936). Chaplin cofounded the modern film production and distribution company called United Artists.

Portable movie theaters, such as this one set up in a rural village in China, gave thousands of people their first sight of events outside their own community. These people are watching a film about everyday life in a large town.

THE LUMIERES

In 1895, two French brothers, Louis Jean Lumière (1864–1948) and Auguste Lumière (1862–1954), set up the world's first movie theater in Paris where audiences paid to watch short films. The films were projected using their own cinématographe, which worked as a projector and portable camera. For several years after this, the Lumières made and showed short films all over the world.

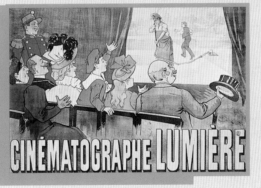

THE
SILENT MOVIES

For the first 30 years of filmmaking, all movies were silent. They had no sound track, so nobody spoke and there were no sound effects. Scenes were introduced by words appearing on the screen. Actors used exaggerated expressions to convey emotions, such as surprise or anger. Each movie theater employed a pianist to play appropriate music to accompany a film.

THE TALKIES

During the 1920s, ways of playing a sound track along with moving pictures were developed. These systems worked well but Hollywood film studios believed that "talkies" (as films with sound were known) were just a novelty. They changed their minds when *The Jazz Singer* (1927), one of the first feature-length talkies released, became a huge box-office hit. The film featured Al Jolson, pictured above.

AN EARLY TELEVISION STUDIO

This picture shows dancers being filmed in a BBC television studio at Alexandra Palace, London, in 1935. Each of the electronic cameras contained a device called an iconoscope. An image of the scene was projected onto the iconoscope by a lens. The iconoscope converted the pattern of light into an electrical signal. The television program would have been shown live because at the time there was no way of recording such programs.

AN EARLY SET

This German television receiver of 1939 is electronic rather than mechanical. The cabinet contains a tall, vase-shaped, glass electron tube. The picture was formed on the wide end of the tube (in the top of the cabinet). The viewer watched a reflection of the screen in the cabinet lid.

HOW TELEVISION WORKS

Televisions can display moving pictures of events as they are happening "live" elsewhere. To achieve this, the light coming from a scene being filmed by a television camera is converted into electrical impulses that travel along wires to a television transmitter. Pictures can then be transmitted to receivers all over the country by using radio signals. To do this, the image is divided into hundreds of lines made up of many thousands of small dots. These are transmitted very rapidly as a radio signal from the transmitter and are picked up by television antennae. The receiver (television set) then reassembles the signal into a picture on the screen. The first experimental television systems were developed in Europe and the U.S. in the 1920s.

Camera collects light from scene.

Camera converts scene into rows of tiny dots. These are then converted into electrical signals.

Electrical signals are sent out from television station to transmitter.

Transmitter converts electrical signals into a radio signal.

Television receiver converts radio signals back into rows of dots on screen.

A television system needs a camera to turn the image of a scene into an electrical signal, and a television receiver to turn the signal back into a picture of the scene. British television pioneer John Logie Baird (1888–1946) was the first person to demonstrate a television system. His camera and receiver were mechanical, with unwieldy spinning discs. But in the U.S., Russian-born Vladimir Zworykin (1889–1982) had already developed the iconoscope (see below). This was the main part of electronic television cameras developed in Britain, the U.S., Germany, and the former Soviet Union, which made Baird's system redundant by the late 1930s.

POOR PICTURES

Baird gave the first public demonstration of a television system in 1926, showing an image of a moving face (right). The picture was made up of just 30 lines and showed ten frames per second, compared to the several hundred lines and 25 or 30 frames per second that make up television pictures today. Although fuzzy and wobbly, the images were amazing at the time.

A MECHANICAL CAMERA

In 1925, John Logie Baird built the first apparatus to take television pictures. At the front of the camera is a Nipkow disc (2) invented by Paul Nipkow. As the disc spins, its lenses focus light from different parts of the subject (1) onto a light-sensitive cell (3). The cell changes the strength of an electric current flowing through it, depending on the brightness of the light that hits it. This is the electrical signal that is sent to the receiver, where it is then turned back into a picture.

BAIRD'S TELEVISOR

The televisor was a mechanical television set devised by Baird. The outside and inside of the machine are shown here. It used a spinning disc with holes that allowed light from a bulb to hit the tiny screen on the right of the televisor. The brightness of the bulb was controlled by a signal transmitted from the television station. Viewers watched the first television broadcasts on these crude sets.

THE SPREAD OF TELEVISION

he first experimental television broadcasts were made in Europe and the U.S. in the early 1930s. The first television broadcasting service was begun by the British Broadcasting Company (BBC) in London, England, in 1936. Only a few hundred people had a television set (which cost as much as a car) and to receive the signals they had to be living near the transmitter. At first, only an hour or two of programs were broadcast each day. Since there was no way of recording them at the time, the programs went out live from a studio. The first broadcast filmed *outside* a studio was in 1937 and showed the coronation of King George VI of England. World War II (1939–45) delayed the development of television in Europe until the late 1940s, but television gained popularity in the U.S. By 1938, there were already 20,000 television sets in service in New York.

ICE-CREAM PARLOR TV

When television was a new medium, it proved a good way of enticing customers into stores to watch and to spend money. Here, children are watching programs in an ice-cream parlor in the 1950s. By 1954, 29 million homes had television sets and by 1960, this had increased to 85 million.

RURAL TV

Because signals can now be transmitted via satellites (see page 27), it is possible to watch television almost anywhere in the world, even in remote areas where mountains would once have blocked the signals. For people living in such areas, television has become a vital source of news, information, education, and entertainment. This picture shows a family watching television in Xinjiang, China.

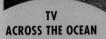

TV
ACROSS THE OCEAN

Telstar, launched in 1962, was the first telecommunications satellite. It relayed the first television signals across the Atlantic. But it was only in the right place over the Atlantic for a few minutes at a time. In 1964, people around the world watched live pictures from the Olympic Games in Tokyo, Japan, relayed via the new geostationary satellites (see page 27).

A SPECIAL EVENT

In Britain, people started buying television sets in large numbers in the late 1940s. By 1950, there were nearly half a million sets in the country. Like this man, many people decided to buy their first television especially to watch the coronation of Queen Elizabeth II in 1952.

HIS MASTER'S PICTURES

This is the famous logo of the Radio Corporation of America (RCA), which developed the first commercial television receiver in the U.S. It also cofounded the National Broadcasting Company (NBC), which gave the first demonstration of television in the U.S. in 1939. At the end of 1940, there were already more than 20 television stations in the U.S.

PICTURES FROM AMERICA

Television networks developed in different countries of the world. At first, programs from one continent could not be seen in another because there was no way of sending the signals so far. In 1962, these live television pictures were the first ones from the U.S. to be seen in Europe. They were transmitted from television stations in the U.S. via the *Telstar* telecommunications satellite.

ADDING COLOR

Originally, all films and television programs were in black and white. Film and television cameras could detect levels of brightness but not different colors. As soon as the basic technologies of film and television were perfected, the next obvious step was to develop systems that could capture and reproduce color. Color film systems became available in the 1930s, despite the resistance of many filmmakers who thought color was just another gimmick. Color television systems were demonstrated as early as 1928, but their use did not become widespread until the late 1960s.

EARLY COLOR TELEVISION

Black-and-white television systems divided the picture into lines of tiny dots and detected the brightness of the light at each one. A color television system detects the brightness of red, blue, and green light at each dot. The television receiver must create the three colors of light, too. This picture shows a color television set from the 1930s.

COLOR BY HAND

The first method of making color films was developed before film could record color. First, the monochrome (black-and-white) film was shot, developed and printed (above). Then each frame of the print was colored by hand with transparent inks (below). This was called tinting. Recently, some old films have been tinted using computers.

KINEMACOLOR

An early attempt at a color film process was Kinemacolor, invented in 1906. It split colors into two parts (rather than the three primary colors), red-orange and blue-green. Frames of each color were recorded alternately on the film, using a rotating color filter to cut out one of the colors on each frame. It worked, but the projected film was of poor quality.

TECHNICOLOR CAMERA

A film shot in Technicolor was expensive. A special camera had one film for each primary color going through it at the same time, so it used three times as much film as a black-and-white film. Technicolor was the most widely used color film until the late 1950s, when film was developed that could record all three colors at once on one film.

PRIMARY COLORS

The three primary colors of light (not pigment) are red, green, and blue. Most colors can be made by combining different amounts of each. Mixing the primary colors in equal amounts makes white light. All film and television systems record color by splitting up the light from a scene into primary colors then using the primary colors to rebuild the original image in color.

TECHNICOLOR

Technicolor, developed in the 1930s, was the first really successful color film process. Colors were divided into the three primary colors, with a separate film for each one. After printing, the films were pressed together to make the full-color film that was projected in the theater.

TESTING THE PICTURE

A test card has grid lines, stripes of different widths, and blocks of color. It is transmitted so that television service engineers can tell when they have the best possible detail and colors on a receiver. Use of test cards is not so important with modern televisions, which are usually self adjusting.

WIDE-SCREEN CINEMA

A standard film (1) is projected onto a flat screen. The Cinerama system (2), introduced in 1952, used three projectors to create a very wide picture on a curved screen. Three cameras, synchronized to take frames at the same time, were needed to shoot the film. CinemaScope (3), introduced in 1953, was the first commercially successful wide-screen system. It requires just one camera and projector. Standard 35-mm film images can be stretched sideways to fill the wide-screen CinemaScope format by using special camera and projector lenses.

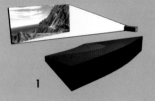

1

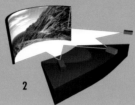

2

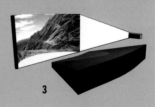

3

FILM SIZES

35-mm film

Film comes in different sizes measured according to its width. Most common is 35-mm film, which is the same as the film used in conventional cameras. Before video cameras came along, amateur filmmakers used 8-mm film. The largest film is 70-mm wide. The larger the film, the more detail that can be recorded on it.

55-mm film

70-mm film

IN THE PICTURE

Large wide-screen pictures, such as the Cinerama picture here, gave the audience the illusion of being in the picture. Films taken from moving vehicles, such as the roller-coaster car, created a very convincing effect.

FILL THE SCREEN

IMAX is the largest film format. The frame is almost square, and is projected onto a screen so big that the audience cannot see anything else but the film. They are fooled into thinking they are actually in the same place as the camera.

Film is a long plastic strip coated in layers of light-sensitive chemicals. Sprocket holes run along either side of the film and are used by cameras and projectors to move the film on from frame to frame. Format is the shape of the picture that appears on the cinema screen and is the same shape as each frame on the film itself. The format used for many years on films and television was called Academy frame. Nowadays, most films have a wide-screen format that is more natural to look at and is better for showing landscapes.

HOW FILM WORKS

Light-sensitive film is exposed to light.

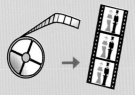

Film   developed using chemicals to make images appear.

Film is projected onto screen.

FILMS IN 3-D

There have been many attempts to create three-dimensional film systems. The most successful were in the 1950s and 1980s, but 3-D films have never become popular because of the special glasses viewers need to wear. Two images, one for each eye, are projected onto the screen. The glasses ensure that each eye sees only one image. In the red-green 3-D system, the two images on screen are green and red respectively. The eye with the green filter sees only the red image, and the eye with the red filter sees only the green image. In this way, a 3-D effect is created.

FILM ANIMATION

nimation is the process of making stationary (inanimate) drawings, 3-D models, and objects appear to move. Animation is used to make cartoons and also to create special effects in other films. To animate a scene, it is photographed many times, with slight changes made for each frame. When projected, this gives the illusion of movement. Originally, all animation was laboriously done by hand, frame by frame, but computers are now used to do most of the repetitive work. For cartoons, illustrators can draw every tenth frame and the computer will automatically fill in a character's movements between them. This is called in-betweening. Similarly, computers can morph (change) one object into another. This technique is used for special effects (see page 19).

COMPUTER MODELS

In films such as *Toy Story* and *Antz* (above), the models of the characters and scenery are not real. They are images stored in a computer. This is called computer modeling. Animators tell the computer how the characters move, and the computer uses its model to draw each frame of the film.

CARTOON PIONEER

Walt Disney (1901–66) began making animated cartoon films in 1923. In 1928, he helped to create Mickey Mouse, possibly the most famous cartoon character of all time. In 1937, Disney released *Snow White and the Seven Dwarfs,* his first feature-length animated film. Sitting with Disney is Donald Duck, who first starred in a film in 1934.

MODEL ANIMATION

Models can be animated in a similar way to cartoons. For each frame, the position of the models in a scene is changed slightly then photographed. This is called stop-action or stop-frame photography. When the film is viewed, the characters appear to be moving. This picture shows a scene from *The Wrong Trousers*, created by Nick Park, which won the 1993 Academy Award for the best animated short film.

CREATE A SIMPLE ANIMATION

You can see how animation works by doing the following simple activity. Hold this book in one hand and flip the pages with the other, watching the penguin image at the bottom right-hand corner. Now make your own simple flip book. Each page of it will represent one frame of a film. You can copy this sequence of cartoon pictures (right), one per page, onto the first 12 right-hand pages of a small notebook, or you can create your own pictures. Be sure to draw each one in the same position on the pages. Animate your pictures by quickly flipping your book.

LIVE ACTION MIXED WITH ANIMATION

In the film *Who Framed Roger Rabbit*, animated cartoon figures were cleverly mixed with live action. While they were being filmed, the actors had to imagine that the animated characters were actually on the set and pretend to talk and react to them (above). The cartoon characters were added later (below).

ANIMATION CELS

Originally, every frame of a cartoon film was drawn individually. The pictures were drawn on thin transparent sheets of acetate called cels and painted in. When all the pictures were complete, they were photographed one by one to make the frames on the film itself.

SPECIAL EFFECTS

Special effects (FX) are the techniques used in film-making and television to create scenes that would otherwise be impossible, dangerous, or too expensive to do in reality. They often include dramatic events that couldn't happen or places that don't exist. Some special effects involve the use of props, such as "glass" bottles made of wax so that they can be broken without harming anyone. Other special effects involve photographic techniques that combine different images to make actors (perhaps filmed in a studio) appear to be moving through a landscape. Many special effects are now created by computer and mixed with live action.

MAGICAL EFFECTS

George Méliès (1861–1938), a French magician and filmmaker, became the first director to make films that told a story. While the Lumière brothers were shooting short films of everyday events, Méliès was using film projection in his stage show, and creating extraordinary fantasy worlds on film. In *A Trip to the Moon*, a 14-minute long film made in 1902, he used scale models mixed with live action to show adventurers traveling to the Moon, meeting aliens, and returning safely to Earth.

COMPUTER EFFECTS

Almost any sort of special effect is now possible with computer graphics. On a computer, backgrounds and objects can be modeled or drawn and even mixed with live action. In *The Mask* (1994), extraordinary distortions of the actors' heads were modeled on computer. Computer effects have taken over from techniques using scale models.

IN THE BACKGROUND

Back projection was a way of providing either a still or moving background for actors working in a studio. Here, the actors are being filmed in a stationary car while previously shot film of a diminishing road scene is projected onto a screen behind them. This gives the impression that they are traveling along the road. Such effects are now created photographically or electronically.

THE BLUE SCREEN

In film and television, blue screen is a method used to mix shots of announcers or actors in a studio with a background from elsewhere. Here, a weather forecaster stands in front of a blue screen (top), while a computer-generated map is electronically added into the image seen by viewers (below). In filmmaking, actors perform in front of a similar blue screen (and often on a blue floor) if their images are to be mixed with filmed or computer-generated backgrounds.

FOOLING OUR SENSES

Good special effects are so convincing that our senses can be completely fooled by them. In this *Back to the Future* theme-park ride, the exhilarating experience of the film is further enhanced by computer-controlled moving seats and sound effects. The two men taking the ride do seem to be reacting as if they are actually in a car flying above city streets.

ON THE SET

Filming is done on a film set. It may be in an indoor studio, on an outdoor set, or on location as shown here. The scene may look chaotic, but everybody is doing a particular job. The camera crew works closely with the lighting crew who operate the powerful floodlight (on the left). At the same time, a sound crew will be recording the actors' voices. It may take a whole day to film what will be edited to become just one minute of the final movie.

TAKE ONE

In filming, a "take" is a continuous shot of a piece of action, such as an actor walking into a room. Several takes of the same shot may be needed before the director is happy that they have got everything right. For reference during editing, details of which shot and take is about to be filmed are written on a clapperboard and filmed by the camera at the start of each take.

STORYBOARDS

A storyboard is a set of drawings made to show how each scene will look when filmed. It helps the director and crew to set up each shot before filming. The picture shows a storyboard for a *Tom & Jerry* cartoon film. Try drawing a storyboard for a scene in a cartoon of your own. Write a simple script and notes about what will happen to the characters in the scene. Think about whether the characters will be in close-up or distant shots and what the background will be.

LIGHTS, CAMERA, ACTION

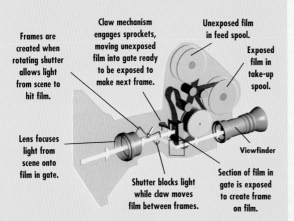

W hen you watch a feature film that is just an hour and a half long, you are seeing the result of thousands of hours of work by hundreds of people. A film starts as an idea, then a script is written to describe the events, the characters, and what they will say and do. The art director must work out what sets and locations will be used and how they will look. Actors are cast for the various parts, and props and costumes must be bought, made, or rented. When all the scenes have been shot, the film is sent to be developed. Prints are viewed by film editors who cut and edit the film, joining the best shots together to make the final version. The sound track of dialogue, sound effects, and music is then added.

BIG REELS

This is an old-style movie camera but it works in the same way as modern ones. On the top of it is a film canister containing two reels, one for unexposed film and the other for exposed film. Such large reels of film are needed because every minute of action uses 30 meters of film. At the front of the camera is a selection of lenses to give different magnifications of a scene.

HOW A MOVIE CAMERA WORKS

A movie camera takes sequential photographs in quick succession and each of them becomes a single frame of the film. As the photographs are taken, the film is continuously fed through the camera, frame by frame, by a mechanism that pulls the film around by its sprocket holes. The lens gathers light from the scene and focuses it onto the chemical surface of the film where it is recorded as a small image. To direct the camera at the scene being filmed, the operator looks through a viewfinder.

Frames are created when rotating shutter allows light from scene to hit film.

Claw mechanism engages sprockets, moving unexposed film into gate ready to be exposed to make next frame.

Unexposed film in feed spool.

Exposed film in take-up spool.

Lens focuses light from scene onto film in gate.

Viewfinder

Shutter blocks light while claw moves film between frames.

Section of film in gate is exposed to create frame on film.

DIRECTION

The director is responsible for how the film looks and sounds on the movie screen. He or she is in charge of all that happens on the film set. Here, the director Steven Spielberg discusses a shot with actor Harrison Ford. The camera is mounted on rails because it will need to move with the action as the shot is filmed.

INSIDE A TELEVISION STATION

A television studio is basically a large, empty room used to film programs as diverse as children's entertainment, pop music, political interviews, discussions, and game shows. Usually, only a small section of the space is used, and the studio set is changed for each program. Most television stations have several studios, and daily programs, such as news reports, occupy a permanent space. Working in the studio are announcers, camera operators, sound engineers, and production staff who control what happens on the set. In a control room outside the studio, program directors and producers control the pictures that are broadcast. Each segment of a program is timed and the various parts carefully coordinated so that they flow smoothly from one to another.

BEHIND THE CAMERA

In this television studio, several cameras are set up to give a choice of views of the program presenter. Powerful lamps in the ceiling light the set. Sounds are picked up by small microphones attached to the announcer's clothing or hanging from above (just out of shot). On the front of each camera is a cue card from which an announcer reads a script while appearing to look directly at the camera.

STUDIO CAMERAS

This camera (left) and its operator's seat are mounted on a counterweighted, swiveling boom so that the camera can film from many different angles. The camera collects light from the scene, splits it into the three primary colors and focuses it onto electronic plates. Electronics measure the brightness and color of each tiny dot of the picture and create an electrical signal that goes to the control room. The camera operator sees the output from the camera on a small viewfinder.

How a television camera works

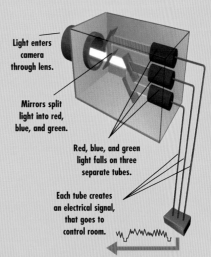

Light enters camera through lens.

Mirrors split light into red, blue, and green.

Red, blue, and green light falls on three separate tubes.

Each tube creates an electrical signal, that goes to control room.

CONTROL ROOM

Pictures from all the studio cameras appear on a bank of screens in the control room. From these, a director selects which pictures are to be broadcast from the station to the viewers. Pictures recorded on video tape, such as television commercials, and from outside broadcasts, such as sporting events, are also displayed on the screens in the control room. These can be slotted into live broadcasts.

FILMING FOR TELEVISION

Most television programs are not broadcast live but are recorded beforehand. They are put together in a similar way as films but are shot with high quality video cameras (see pages 28–29). This commercial is being shot on a studio set. The video will be edited afterwards and perhaps special effects added before the final version is ready to broadcast.

MAKING PICTURES APPEAR

To display a moving picture, a device is needed that will show each frame for a split second before moving on to the next one. Film is shown by projecting light through a film and onto a large screen where a magnified version of the picture appears. A television receiver (television set) is controlled by the signal it receives from a transmitter. Most receivers contain an electron tube with a glass screen. Hand-held televisions and flat-screen televisions have displays that show pictures using liquid-crystal display (LCD).

PROJECTING AN IMAGE

Film is shown by using a projector. A projector does the opposite job of a film camera. Each of the thousands of frames of a film passes briefly in front of the projector's powerful light source, while the lens focuses a greatly magnified version of each image onto a screen.

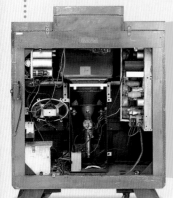

AN OLD TV SET

With the back removed from this first-generation television set, its tube is visible beneath a small screen. It was positioned vertically because the tube was so long. There were no semiconductors at the time, so the electronic components are much larger than they are in today's sets. It took a minute or more for them to warm up to operating temperature.

LCD DISPLAY

The screen of a hand-held television works in the same way as a calculator display. Tiny dots of liquid crystals allow light through or block it to change the brightness. Colored filters make the dots red, green, and blue. The dots make up the picture in the same way as rows of dots on a television tube (see far right).

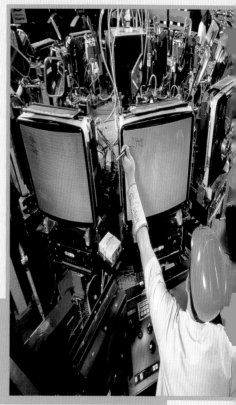

MAKING TUBES

At electronics factories, television tubes have several parts installed in them before being closed and sealed. The air is then pumped out to form a vacuum inside the tube. Each tube is tested before being installed in a television set. Improving technology has produced tubes with large, square, flat screens.

LOOKING INSIDE A TELEVISION

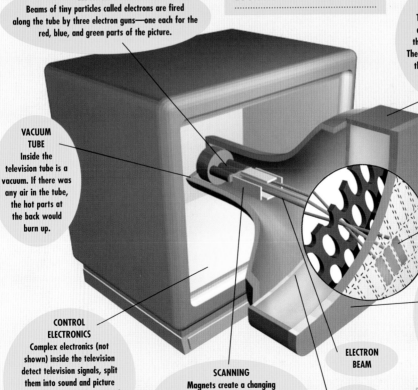

ELECTRON GUNS
Beams of tiny particles called electrons are fired along the tube by three electron guns—one each for the red, blue, and green parts of the picture.

TELEVISION TUBE
The main part behind a television screen is the glass electron tube. The base of the tube forms the screen at the front of the television.

VACUUM TUBE
Inside the television tube is a vacuum. If there was any air in the tube, the hot parts at the back would burn up.

SCREEN CLOSE-UP
Enlarged section showing the mask, electron beams, and red, green, and blue dots on screen.

CONTROL ELECTRONICS
Complex electronics (not shown) inside the television detect television signals, split them into sound and picture signals, and control the strength and movement of the electron beams that rebuild the picture.

SCANNING
Magnets create a changing magnetic field that makes the beams scan across the back of the screen in a zig-zag pattern, rebuilding the picture line by line.

ELECTRON BEAM

MASK
Allow beams to hit correct color dots.

TV SCREEN
Where the electrons hit the back of a screen, colored dots of a chemical called phosphor make light that is seen by the viewer.

25

SATELLITE DISHES

A satellite dish is an antenna that sends or receives signals from communication satellites high above the Earth. This ground station receives signals beamed at it from a satellite. The signals could be television pictures on their way from one television station to another, or telephone calls.

BROADCASTING WITH RADIO SIGNALS

In traditional broadcasting, electrical signals are sent to an antenna called a transmitter, where they are converted into a radio signal. The signal spreads in all directions from the transmitter and can be detected by the antenna of any receiver within range. Transmitters are placed on top of tall masts so that the signals can spread as far as possible in all directions. Eventually, the signals become too weak to receive, so many transmitters are needed to cover a wide area. The signals can be in analog or digital form.

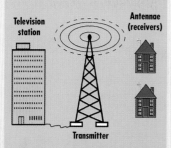

Television station

Antennae (receivers)

Transmitter

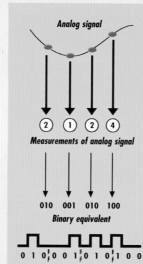

Analog signal

② ① ② ④

Measurements of analog signal

010   001   010   100

Binary equivalent

0 1 0 , 0 0 1 , 0 1 0 , 1 0 0

Digital signal
1 = signal   0 = no signal

ANALOG & DIGITAL BROADCASTING

An analog signal changes strength all the time, and is shown in diagrams by a wavy line. Here you can see how an analog signal is changed into a digital signal, which only has two different strengths. Each 1 creates a signal, while a 0 produces no signal. If an analog signal is distorted as it travels, the quality of the picture it is carrying is affected, whereas a distorted digital signal is still clear enough to create a perfect picture.

TUNING IN

Television signals for programs on different channels and from different television stations are transmitted using different radio waves or electric currents called carriers. When you tune a television receiver, you are selecting the carrier of the station you want to watch. Here, the same signal is being sent on two carriers of different wavelengths.

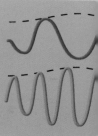

Television signals can travel from a television station to your home in several different ways. The original method of using radio signals is still in use (see below). With cable television, the signals are transmitted along underground cables. With satellite television, the signals travel as radio signals to and from a satellite orbiting the Earth. Television pictures can also be sent over the Internet, a process known as Webcasting. The latest major development in television is digital broadcasting. Digital signals travel by the same routes as normal signals (called analog signals), but digital broadcasting allows many more channels to be broadcast at once, and gives much clearer pictures.

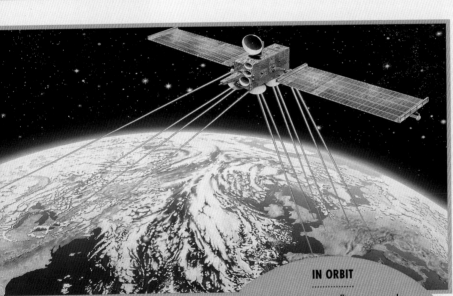

NOT FOR BROADCAST

In a closed-circuit television (CCTV) system, pictures from cameras go directly to receivers that are seen by just one or two people. The most common use of closed-circuit television is security. Here, a street surveillance camera has recorded someone stealing a handbag. The recording will be used to try to identify the thief, who might be prosecuted later.

IN ORBIT

A communications satellite acts as a relay station in space. It collects signals beamed up from ground stations and retransmits them to other ground stations. Communications satellites, including those for domestic satellite television, are in a special type of orbit called a geostationary orbit, which means they always stay above the same place on the Earth's surface as the Earth spins. Signals from satellites reach receivers across a much greater area than those from transmitters on the surface.

CARRIER SIGNALS

The frequency of the bottom carrier signal is higher than in the top signal, but both carry the same information.

Before the arrival of video, film was the only way of recording pictures. On television, everything was live, except for films. These were televised by a special device that was part projector, part camera. To store a program shot by television cameras, it is necessary to record the television signals. In a video machine, the signals are stored as a pattern of tiny magnetic particles in the magnetic layer on a tape. Once recorded, the programs can be played back at any time. In digital video, the signals are stored in the form of numbers, either on tape or as computer files. With the appropriate equipment, they can be played on a television or a computer, and even over the Internet.

VIDEO HISTORY

Magnetic video taping was first demonstrated in 1951, but it was 1956 before television programs were recorded in this way. The early video recorders were big, expensive machines used only by broadcasters. Compact video recorders that used video cassettes became available to the public in the 1970s.

HOW VIDEO WORKS

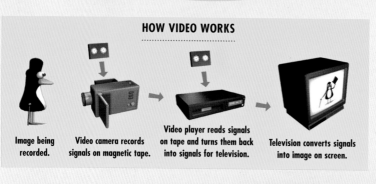

Image being recorded.

Video camera records signals on magnetic tape.

Video player reads signals on tape and turns them back into signals for television.

Television converts signals into image on screen.

VIDEO AT HOME

Standard domestic video recorders use magnetic tape to record television programs. These machines often include an automatic timer that allows programs to be recorded and then watched when the viewer wants. Videotape is likely to be replaced by the digital video disc (DVD), together with DVD players (right). DVD gives better picture quality, with almost instant access to any part of a recorded program.

VIDEO EDITING

Video is edited in a similar way as film to create a finished program. The required sections of a videotape are copied electronically onto a fresh tape, then special effects and sound tracks are added. Most editing is now done in digital editing studios after the videotapes have been digitized and stored on computer. The completed video can then be copied any number of times onto other tapes for use.

Mini video cassette.

ELECTRONIC FILM

A hand-held video camera records moving pictures onto videotapes that can be played back on television. The camcorder illustration (right) shows how the lens at the front collects light from a scene and focuses it onto a special microchip called a charged coupled device (CCD). The CCD detects the color and brightness of the light hitting it and converts this information into a signal for the tape. Increasingly common are digital video recorders on which video is stored in memory inside the machine.

Charged coupled device (CCD).

Viewfinder with small LCD screen.

PLAYING DVD

A digital video disc (DVD) is similar to a CD-ROM, but can hold far more information, such as the huge amount of data needed to display a feature-length movie. Portable DVD players are available with small screens or headsets with screens in them. This makes it possible to watch films almost anywhere.

CHANGES WITH TECHNOLOGY

As the technology used to make and show films and television programs has advanced, it has affected the way films and programs look and how we watch them. The changes that have taken place in television have been greater than those in film, in which the use of special effects has been the main development. The look of many television programs, especially those for young people, reflects the design and fashion ideas of the time. In the 1950s, people would have found programs made up of short, jumpy clips very strange to watch, but this technique is very popular today. Television provides far more now than in the past. We have come to expect many channels—providing news, entertainment, sports, education, information—and to be able to watch television virtually around the clock.

KEEPING UP WITH THE CHASE

The first action movies were comedies, such as those featuring the famous Keystone Cops, which involved police chasing villains. The films were shown at an increased frame rate to make the action appear faster. Even so, by the standards of today's action movies, they look very slow. Modern action movies are full of high-speed events and special effects that look far more convincing.

COURT-SIDE SEATS

Television coverage of sports has changed greatly. Digital television allows the user to pick which camera angle they want to see and to ask for instant replays. At the events themselves, referees can watch replays provided by the television companies to help them with difficult decisions, such as whether a ball crossed a goal line or not. The crowd watches replays on giant video screens. This is a high-definition camera that can record pictures in much greater detail than a normal camera. Pictures are very sharp when viewed on a high-definition television (HDTV) set.

HOME THEATER

There has always been competition between movies and television as forms of entertainment. The technology of television now allows people to have huge television screens, with the ability to show widescreen formats, with CD-quality sound. With the addition of a video or DVD player, systems like these have become known as home theater. In the future, films may be released via the Internet and viewed from home theaters.

INTERACTIVE TELEVISION

In standard television, the signals travel one way only, from the television station to the receiver. In interactive television, the viewer can send information back to the television station along a cable or telephone line. For example, a viewer can choose items to buy on a shopping channel simply by selecting them on the screen. The latest television sets will act as Web browsers as well as televisions.

OLD & NEW

By comparing a television set made in about 1950 (right) with what is possible today (left), you can clearly see how much technology has changed the way that televisions look. These changes have come about through miniaturization of electronics and the development of liquid-crystal displays. In the future, hand-held screens (left) may become a common part of people's personal equipment, used for phone calls, messages, and entertainment.

GLOSSARY

Analog signal
A type of signal used for broadcasting. Analog signals continually vary their strength and often reduce the picture quality when they become distorted.

Digital signal
A type of signal used for broadcasting. Digital signals have only two different strengths, and are still clear enough to create a perfect picture even when they become distorted.

Electrical signal
An electric current that changes in strength and direction. The changes represent some sort of information, such as the pattern of light in a television picture, or the shape of a sound.

First-generation
The first type of a device that is made. For example, first-generation television sets had small black-and-white screens in a huge cabinet.

Microchip
A slice of semiconductor material, often just a few millimeters across, which contains hundreds or thousands of microscopic electronic components. Most microchips are designed to do a certain job, such as amplifying (increasing) an electrical signal.

Semi-conductor
A type of electronic material that can be a good conductor or bad conductor of electricity. Electronic components such as transistors are made from semi-conductors.

Television camera
A camera that turns an image into an electrical television signal.

Television receiver
A device that turns an electrical television signal into a moving image on a screen.

Transmitter
An antenna that sends out radio waves.

Wide-screen
Describes a film or television picture that is much wider than it is tall.

ACKNOWLEDGMENTS

We would like to thank David Rooney and Elizabeth Wiggans for their assistance. Artwork by John Alston
First edition for the United States, its territories, dependencies, Canada, and the Philippine Republic, published 2000 by Barron's Educational Series, Inc.
Original edition copyright © 2000 by *ticktock* Publishing, Ltd.   U.S. edition copyright © 2000 by Barron's Educational Series, Inc.
All rights reserved. No part of this book may be reproduced in any form, by photostat, microfilm, xerography, or any other means, or incorporated into any information retrieval system, electronic or mechanical, without the written permission of the copyright owner.
All inquiries should be addressed to: Barron's Educational Series, Inc., 250 Wireless Boulevard, Hauppauge, NY 11788 • http://www.barronseduc.com
Library of Congess Catalog Card No. 99-67204   International Standard Book No. 0-7641-1065-9
9 8 7 6 5 4 3 2   Printed in Malaysia
Picture research by Image Select.

Picture Credits: t = top, b = bottom, c = center, l = left, r = right, OFC = outside front cover, OBC = outside back cover, IFC = inside front cover

AKG; 8c, 31br. Ann Ronan @ Image Select; 4/5t, 5r, 9br. Corbis; 11cl. Hulton Getty; 6cr, 9bl, 9tr, 10tl, 10/11b, 10/11t, 16b, 18cl, 28cl. Image Bank; 12/13 (main). Image Select; 11 (main), 14tl, 14cl, 14bl, 17cl, 17cr, 21br. Image Select/BFI; 7c. JVC (UK); 28br. The Kobal Collection; 7r, 18/19 (main), 30cl. Mary Evans Picture Library; 5b, 6bl, 8/9tl. NMPFT/Science & Society Picture Library; 4l, 12br (main), 24cl. PIX; OFC (main), OFCcl. Rex Features; IFC, OBC, OFCtl, OFCbl, 2br, 2l, 13b, 15r, 16/17t, 18/19b, 12bl, 12tl, 19br, 19cr, 20c, 20r, 20/21t, 23br, 29cr, 30/31b, 30/31c, 30/31t. Science Photo Library; 3bl (Oscar Burriel), 12tr (J.L. Charmet), 22/23 (main), 28/29 (main, C.S. Langlois), 30bl (Philippe Plailly), 32tr. (Oscar Burriel). Telegraph Colour Library; OFCbr, OFCcr. The Ronald Grant Archive; 6/7bc, 6/7t (© Barry Norman), 12c, 14br, 14/15t, 16/17 (main), 17b, 19t. The Stockmarket; 22bl, 23cl (John Madere). Tony Stone Images; 3cr, 10cr, 21tr, 24/25b, 24/25t, 26/27c, 26/27t, 27r. Zefa-Stockmarket; 2/3t.

Every effort has been made to trace the copyright holders and we apologize in advance for any unintentional omissions.
We would be pleased to insert the appropriate acknowledgement in any subsequent edition of this publication.

BARRON'S